Protecting Earth's Resources

Reader

Core Knowledge®

Printed in Canada

ISBN: 978-1-68380-550-2

Protecting Earth's Resources

Table of Contents

Water Resources

Chapter 1

Look under a sink, and you will see pipes. Fresh water goes to and from the sink through these pipes. Water comes into a building through pipes. It comes out through the faucet of the sink. Water goes down the drain in the sink. It then leaves the building through different pipes. What happens to water after it washes down the drain?

Big Question

What is water quality?

Hot water and cold water flow to the faucet through the smaller pipes. When water goes down the drain, it leaves the building through the larger pipes. Where does it go next?

Water Use Produces Wastewater

After people use water, we call it **wastewater**. Wastewater might contain dirt, germs, or harmful chemicals. People and animals could get sick if they drink it. Wastewater put back into rivers, lakes, and oceans should be cleaned first. Several steps are needed to clean the dirty water. Only then is it safe to put back into the environment.

Vocabulary

wastewater, n. water that has been used by people

water treatment plant, n. a facility that cleans wastewater before it is discharged into the environment

Wastewater is cleaned at special sites. These are **water treatment plants**. If you live in a city, you might have one of these sites near you. Different structures filter the water. The process uses filters and chemicals that kill germs. The clean water is discharged, or released, into lakes, streams, or the ocean. From there, it can be used again.

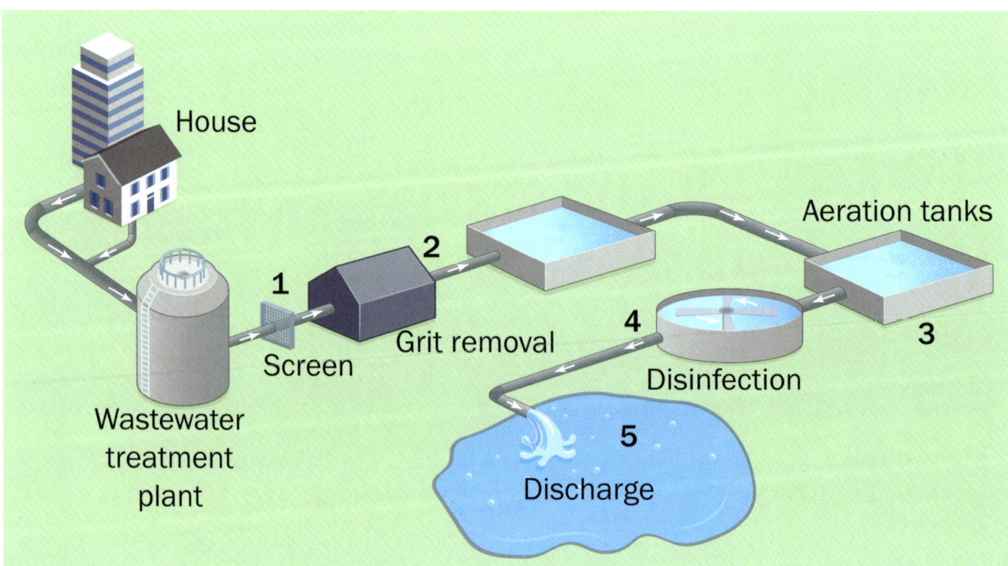

Wastewater Is Treated and Reused

The water that leaves your house and other buildings flows through pipes. This wastewater flows into sewers. Sewers are large pipes. Water from many buildings flows through sewers.

1. Water from sewers reaches the plant. Here, it first flows through screens. The screens filter the water. They remove large objects from the wastewater.
2. Next, water flows to the grit removal chamber. In the chamber, water spins quickly. The spinning removes sand, dirt, and other tiny bits of matter from the water.
3. Then, the water flows into tanks. Air and **bacteria** are added to the water in the tanks, making it bubble. Bacteria are too small to see without a microscope. But they do a big job! They break down any material left in wastewater after the first two steps.
4. The last step in the cleaning process is **disinfection**. Look at the word: *disinfection*. You can find the word *infect*. The prefix *dis-* in front of a word means "the opposite." *Disinfect* means the opposite of *infect*. In this process, workers add chemicals to the water that kill the bacteria. The chemicals also kill other germs in the water that could make people sick.
5. After this, the water is safe to use again. It may go through more filters so it is safe to drink. It may be returned to the environment through streams and other bodies of water.

Vocabulary

bacteria, n. tiny organisms that break down matter

disinfection, n. the process of cleaning infection from a material

Wastewater Contains Chemicals

Wastewater sometimes contains materials that water treatment plants cannot remove. These materials may stay in water after it is treated and can cause **problems** in the environment.

> **Vocabulary**
>
> **problem, n.** a want or a need

Surfactants: Recall when you last washed clothes or your hair. Did you notice a lot of bubbles? Certain chemicals cause these bubbles. We call such chemicals surfactants. Surfactants are in most soaps. They are in dish, laundry, and hand soap and shampoo. We use a lot of these chemicals in our cleaning products. Surfactants stick to dirt on clothes or hair. They help to remove dirt. These chemicals then wash down the drain. It is hard to clean them from the water.

Chemicals in water can have effects. How are the chemicals in this water affecting the water?

Water containing surfactants can end up in lakes, rivers, or the ocean. Surfactants can be a problem because they may harm wildlife. The chemicals stick to the gills of fish. The fish cannot get oxygen from the water as well. Scientists noticed the effects of these chemicals on fish. They asked companies to stop putting surfactants in cleaning products.

Phosphates: How do we know that treated water is safe? Scientists do **water quality testing**. The tests measure the chemicals that could be harmful in water. Water quality testing is first done before putting water back into lakes or rivers. Scientists also test water in lakes and rivers. They want to check that it is safe for wildlife.

Scientists test for a chemical called phosphate. Phosphate dissolves into water from soaps, such as dish and laundry soap. It also can enter into water from fertilizers. Farmers and gardeners use phosphate. They use it to help plants grow. But plants may not use all the phosphate. Too much phosphate in water causes algae to grow out of control.

Vocabulary

water quality testing, n. methods to measure the types and amounts of chemicals in water

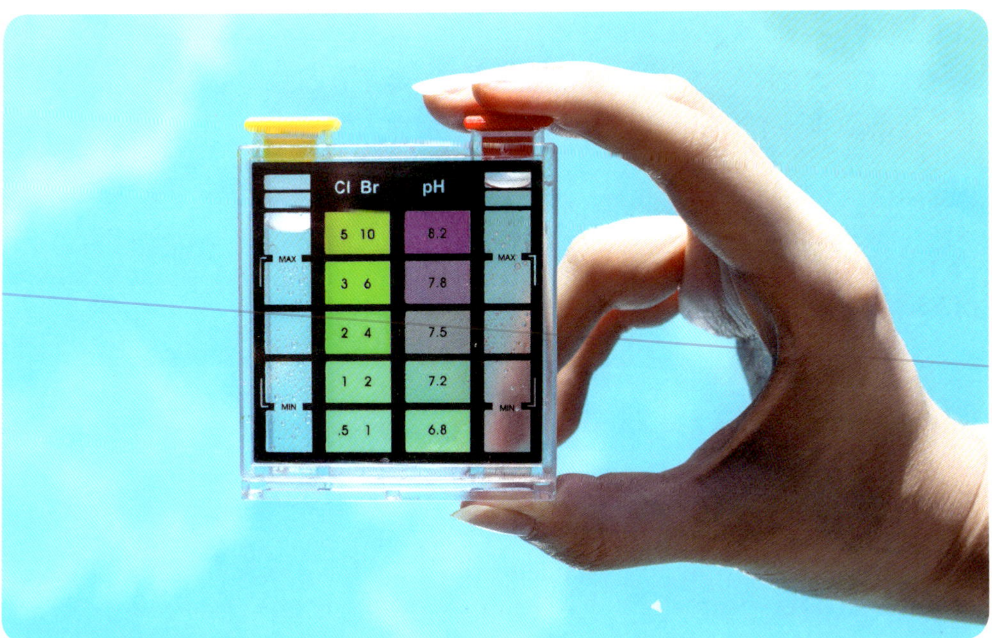

Scientists use water quality testing to measure the kinds and amounts of chemicals in water. By testing water, we can know it is safe to swim in, drink, or discharge from a water treatment plant.

Too Much Algae Is a Problem

Algae are photosynthetic organisms that grow in water. They are part of the food chain. However, algae can be harmful if they grow out of control. When algae reproduce uncontrollably, we call it an **algal bloom**. In this case, the algae can cover the surface of water.

Vocabulary

algal bloom, n. rapid growth of microscopic algae in a body of water

Blooms are harmful because they reduce the amount of oxygen in water. Algae in the bloom have a short life cycle. They grow and die every day. When they die, bacteria flourish and soak up oxygen from the water. As the oxygen is used up, other organisms such as fish cannot survive. Then, as the fish die, more bacteria use up yet more oxygen. In the worst cases, the lake ceases to be a good habitat for many living things.

Phosphate can cause algal blooms to cover ponds and waterways. Algal blooms are harmful to plants and animals. Look at the labels of cleaning products at your house to see if they are phosphate free. Limiting phosphate use can prevent algal blooms.

Because of this harm, scientists have asked companies not to put phosphate in their products. This has helped to decrease algal blooms. However, phosphate is such a useful product that it is still widely used. Farmers and gardeners can help by using less phosphate. You can help by using cleaning products that do not have phosphate.

Water Resources, Problems, and Solutions

Chapter 2

When astronauts first looked at Earth from space, they said it looked like a "blue marble." Why does Earth look blue? Water! Without water, there would be no life on Earth. When scientists examine other planets, such as Mars, they look for signs of water. Water is one of the most vital resources that humans need to live. If water can be found on other planets, there is an increased chance that people could live on planets besides Earth.

Big Question

How can human activities affect the quality of water?

Why is water such an important resource? Plants and animals live in water, drink water, and use water to grow and make food. No living thing can survive without water. Therefore, people must work hard to protect the water we have on Earth.

Most of Earth's surface is covered by water. Can you see any evidence of water on Mars in the photo? Why is water necessary for life?

Earth's Fresh Water Is Limited

About three-quarters of Earth's surface is covered in water. However, only about 1.5 percent of the water on Earth is considered **fresh water**. Some fresh water is the frozen water of ice caps or glaciers.

Most fresh water is in rivers, lakes, and streams. This water is on Earth's surface. We call it **surface water**, and it comes from rain.

Other fresh water is found underground. It is trapped between layers of rock in spaces called aquifers. This water is known as **groundwater**. Groundwater can rise to the surface in springs. Groundwater can also be pumped up to the surface from underground. Most groundwater is brought to the surface by wells. People have used wells for thousands of years to access groundwater.

Vocabulary

fresh water, n. water that is usable by humans for things such as drinking and cleaning

surface water, n. fresh water found in rivers, streams, and lakes on Earth's surface

groundwater, n. water stored in the spaces between materials beneath Earth's surface

Where do you think the water in this spring came from?

Fresh Water Is a Renewable Resource

Surface water and groundwater are replaced by rainfall. Rainwater runs off into streams, rivers, and lakes. It also seeps underground, where it enters an aquifer. Resources that are quickly replaced are known as renewable resources. When something is renewable, it can be naturally replaced, or replenished.

But what happens if we use up renewable resources faster than they can be replaced? These resources are limited resources. Fresh water is replenished by rain, but people around the world are concerned about using fresh water faster than it is replenished.

In the United States, people's water use is monitored by the Environmental Protection Agency as well as by state and local agencies. If fresh water gets used faster than it can be replaced, these agencies may suggest ways or pass laws to help **conserve** the water. For example, a government agency may not allow people to water their lawns. There are many ways that people can conserve water.

Vocabulary

conserve, v. to save or protect

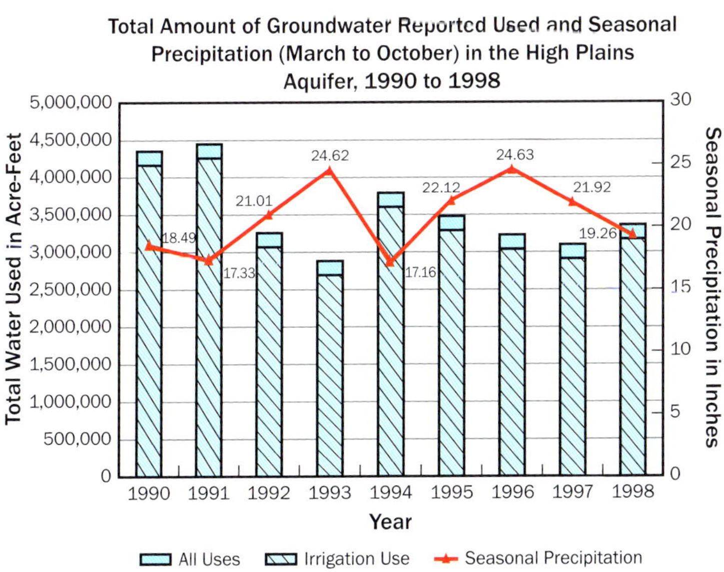

This graph shows that during some years there is not enough rainfall to replace the water that people use.

Harmful Chemicals Cause Water Pollution

Conservation is one way to take care of water resources. Another way to care for water resources is to prevent water pollution. **Water pollution** occurs when harmful or poisonous materials enter into water. When water is polluted, it may harm the living things that depend on it for survival.

> **Vocabulary**
>
> **water pollution, n.** the presence of harmful substances or matter in water
>
> **pollutant, n.** an artificial or natural substance that contaminates air, water, or soil

Chemicals can pollute water. For example, chemicals from your home, such as detergent and shampoo, wash down the drain. But homes are not the only place where chemicals enter the water supply. There are many industries, such as agriculture, that contribute to water pollution.

Some farmers use chemicals on crops to help them thrive. They use chemicals to keep weeds from growing among crops. They also use chemicals to keep insects from eating the crops and damaging them. All these types of chemicals can wash off into nearby surface water when it rains or seep below ground into the groundwater, becoming **pollutants** in our water supply.

Farmers often use chemicals on their crops. What are some advantages to using chemicals on crops? What are some disadvantages?

Oil Spills Cause Environmental Damage

Oil is a mixture of chemicals. It is an important **natural resource** because people depend on it for energy. Oil companies access oil in Earth by drilling underground and pumping it to the surface. The oil is then transported to where it is needed. Sometimes, this oil spills into water when companies are drilling, pumping, or transporting it. When oil spills into water, it can harm the living things in the aquatic ecosystem. For instance, during an oil spill, small algae, called plankton, can die. Fish cannot get oxygen from the water. Even birds cannot fly with oily wings.

Vocabulary

natural resource, n. materials that occur in nature that can be used by people

It is very difficult to remove oil from water once it has been spilled. Scientists can try to remove it by soaking it up. They have also tried to remove oil by releasing bacteria or chemicals into the area of an oil spill that break down the oil. Scientists can skim oil that has spilled from the water's surface. Scientists can also burn oil that has spilled on the surface of water. Although these ways to remove oil can help, It is very difficult to remove all the oil from water once it is there.

Scientists skim oil from the surface of the water after it has spilled. This boat uses a boom, the orange tube, which rides on the surface of the water, to collect spilled oil.

The Pacific Garbage Patch Is a Symbol of Damage

There is a new island in the Pacific Ocean. It is not one that people would want to visit because it is made of trash. How did the trash get there?

Wind and rivers can move plastic into the ocean. People can throw garbage into the ocean. Once garbage gets into the ocean, currents carry it away. Over many years, the garbage ends up where the currents meet. This is the area known as the Pacific Garbage Patch.

Sunlight and water can break down plastic into small bits. They remain in the ocean, even if they are hard to see. The plastic is hard to remove from the water without taking tiny plants and animals out, too. How can all this pollution be removed from the ocean?

One young man, Boyan Slat, developed a way to remove this plastic. His invention uses floating booms to gather the plastic into a holding tank. Ships empty the tanks and bring the plastic to a recycling center. Scientists are not sure if the invention will be enough. In any case, people need to figure out how to solve the problem of the Pacific Garbage Patch.

Boyan Slat developed an idea to remove pollution from water. He designed and built his invention with the aim of removing thousands of tons of garbage from the Pacific Ocean.

People Can Respond to Environmental Damage

We are taught to not litter and to try to keep our neighborhoods nice. But trash does collect in many public areas. Used paper cups, water bottles, aluminum cans, and plastic wrappers find their way into our parks, beaches, streets, and sidewalks and around trees and buildings.

Trash builds up in neighborhoods, especially after community events.

Cleaning up and removing waste materials can be part of living or visiting an area. Many cities have clean-up days. On these days, people are invited to help clean up the pollution in their community.

You do not need to wait for a clean-up day to pick up trash that you see. Try living by the rule to leave the environment better than you found it when you are outside.

Helpful notices encourage people to look after the environment. What do you think this sign at a park is meant to do?

Germs in Water Can Be Harmful

Even when water looks clean, it could still contain pathogens. Pathogens are germs that cause illness or make you sick. They are living things that are too small to see with your unaided eyes. Many tiny things living in water will not cause harm. But some germs in water can make people extremely sick.

A giardia infection can spread quickly from person to person in places where water is unsafe. It can be picked up from all kinds of water sources, including lakes, streams, swimming pools, city water, and wells. Even washing fruits and vegetables in contaminated water can make you sick.

It is not safe to drink water that has not been treated. Water in your home may come from a river. It may be pumped from underground. Either way, the water that you get in your home has been filtered, tested, and made safe for you to drink.

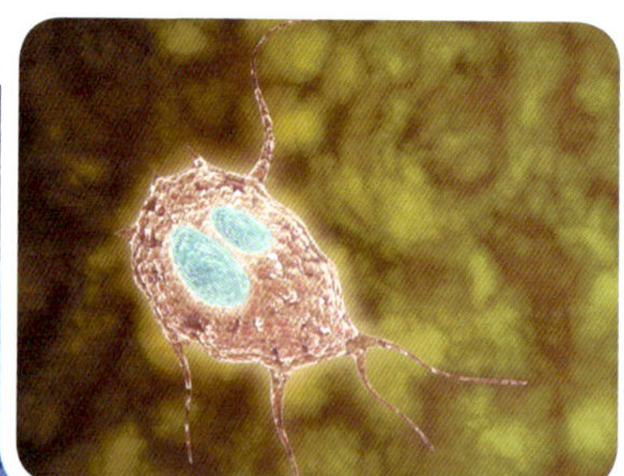

Microorganisms such as this can make water unsafe to drink.

This water looks good enough to drink. It could have tiny living things in it that pollute the water and can make people and animals sick. Never drink water that has not been treated.

Air Pollution

Chapter 3

Can you imagine wearing a mask to filter the air you breathe when you go outside? In some parts of the world, there is a large amount of **air pollution**. Air pollution is any type of material that is in air and makes it unsafe to breathe. Air masks can filter pollution from the air so that people do not breathe in toxic gases, fumes, or small particles.

Big Question

What is air pollution?

Vocabulary

air pollution, n. any type of substance that is in air and makes it unsafe to breathe

People who live in cities with a lot of air pollution must wear air masks to enjoy outdoor activities.

Smog Is a Problem

Have you ever watched a piece of wood burn in a campfire or fireplace? When materials such as wood burn, smoke is given off. This smoke rises into the air.

In large cities, a lot of smoke gets released into the air. This is because cars and trucks burn fuel and release clouds of smoke. Factories and power plants may burn oil, coal, or natural gas to get energy. They have tall stacks that release ash, smoke, and chemicals into the air.

In crowded cities, smoke from burning fuels sticks to particles of water in the air. When the mixture of smoke and air is heated by sunlight, the result is **smog**. The word *smog* comes from two words, *smoke* and *fog*. Smog makes air look hazy. Smog can cause health problems for people.

Vocabulary

smog, n. a type of air pollution that results from a mixture of gases from burning fuels, air, and sunlight that makes air look hazy

Smog is hanging in the air over this city. On days with a lot of smog, people may be warned to stay indoors.

Effects of Smog Are Harmful

Your lungs take in air and use the oxygen from it. If other materials are in the air, your lungs can become irritated. People may cough or have trouble breathing. Air pollution can also irritate your eyes by causing them to become red or itchy or to burn. In some areas of the world, especially large cities with areas that manufacture goods, smog causes major health problems. How are these places dealing with the problem of air pollution?

Smog can be a serious problem for people with asthma. Asthma is a condition that makes breathing difficult. If people with asthma live in or visit an area with a lot of air pollution, they could have serious trouble breathing.

Plants Can Help Solve the Problem of Air Pollution

Plants and Air Pollution: In addition to taking in carbon dioxide and releasing oxygen, plants absorb many air pollutants. Plants such as daisies, English ivy, and evergreens trap air particles that cause pollution and deal with air toxins. Using plants to clean the air does not require any special equipment because the plants do the work.

Urban Gardens: Adding plants in cities helps clean the air. But where can plants be added in space that is crowded with buildings? One city in Italy came up with a creative way to help address the issue of polluted air. A famous designer built these tall apartment buildings covered by a vertical garden.

This photo shows apartment towers in Italy covered in vertical gardens. The designer of these towers has built vertical garden apartments in China and Canada also.

The designer of the apartment buildings came up with the idea for a vertical garden because plants can be used to help clean the air. There are five hundred medium and large trees, three hundred small trees, five thousand shrubs, and eleven thousand other plants growing on the outside of these apartment buildings, all of which are cared for by arborists. An arborist is a person who knows how to keep many types of trees and other plants healthy.

Recall that plants use sunlight and carbon dioxide from the air to make food. During this process, plants release oxygen. The plants growing on these apartment buildings will turn 44,000 pounds of carbon dioxide into oxygen in just one year alone! In addition to removing carbon dioxide from the air, trees also help filter dust particles, making air cleaner.

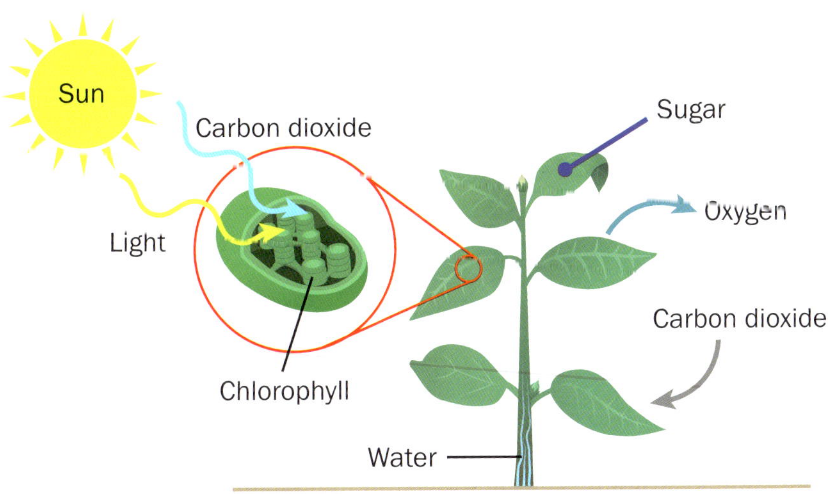

Another benefit to this vertical garden is that the plants provide shade to keep temperatures cool inside the apartments. People who live in these apartment units will save money and energy on cooling their homes. The leaves of the plants can also help absorb noise pollution from the city below.

Engineering to Improve Air Quality

Other cities with high amounts of air pollution have been finding ways to use plants to clean the air. In Singapore, designers have built "supertrees."

Look closely at the "supertree." It is made of 150,000 plants. The structure that the plants grow on collects energy from the sun. Each Supertree provides light for the city at night.

In Bangkok, Thailand, pillars along the highway are covered in plants. These highway pillars use plants in unusual spaces to reduce air pollution.

What is the advantage to covering pillars along the highway in plants?

The Need for Clean Air

Chapter 4

Big Question

How can human activity affect the quality of air?

Animals on Earth cannot live without oxygen. Even animals that live in the sea need oxygen! Some get oxygen by filtering oxygen gas that is dissolved in water with their gills. Others get it by coming up to the surface for a breath. When animals breathe air, their bodies use the oxygen in it. If the air that animals breathe has dirt and chemicals in it, the animals can get sick.

External gills

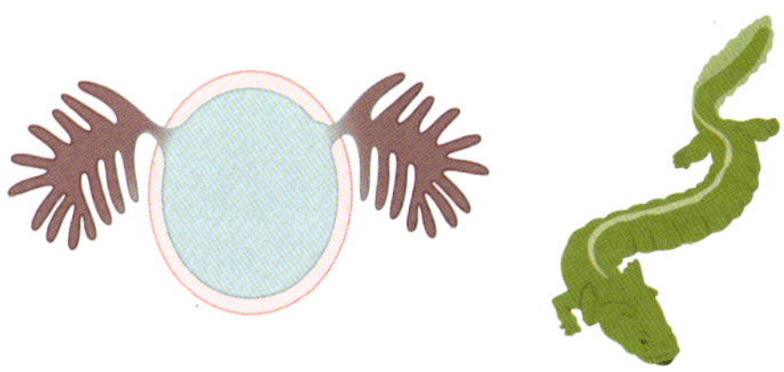

Internal gills

Lungs

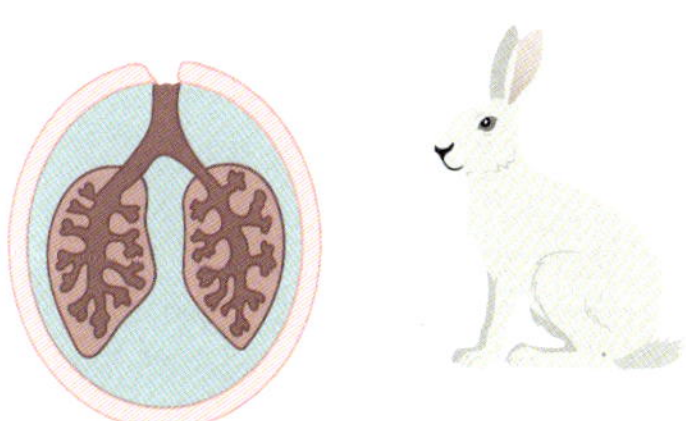

Tracheae

These diagrams show different structures organisms have for taking oxygen from their environments.

Air Quality Is Important

People know the air is polluted when we have trouble breathing or seeing. To find out the degree of air pollution, the Environmental Protection Agency is responsible for measuring air quality. **Air quality** is a measure of how clean the air is. Even if air looks clean, it can still have pollutants in it, particles that do not belong in the air. What pollutants are found in the air, and how do they get there?

Vocabulary

air quality, n. a measure of how clean the air is

These are photos of the Chrysler building in New York City. Smog makes the air look hazy or dirty. When the sky is hazy, plants do not grow as well. Can you explain why?

Smoke: Smoke and ash are released into the air anytime materials are burned. Forest fires, smoking cigars or cigarettes, campfires, car and truck exhaust fumes, and factories all create smoke. Smoke is a common air pollutant. When it mixes with water vapor in the atmosphere, it forms smog. This gives the sky a hazy appearance.

Ozone: **Ozone** is a gas found in the air. Ozone protects us from ultraviolet light released by the sun. It does this by absorbing dangerous sunrays and preventing sunburns. However, when ozone is near the ground, it is harmful because we can inhale it. Ozone mixes with smog to make the sky appear hazy on warm summer days. Ozone in the air makes it hard to breathe.

Vocabulary

ozone, n. a gas found in air

Dirt, Dust, and Sand: Plants are important for more than just cleaning the air. They also hold on to soil and sand with their roots, trapping the soil and sand so that it does not blow away or blow into our air.

In some windy places, dirt, dust, and sand can become pollutants. In the early 1930s, people did not protect the soil as much as we do now. Loose soil blew into the air, blocking out the sun. Plants and animals could not survive these conditions, and the dirt in the air caused many health problems across a large area.

Air Quality Is Measurable

The Environmental Protection Agency measures the amount of each type of pollutant in the air and rates how clean it is using the Air Quality Index. The Air Quality Index is a system of colors and numbers. The Air Quality Index lets people know the air quality in their region daily. By looking at the color and number, people can know how safe or polluted the air is on any given day. The lower the number, the cleaner the air is. If the number is higher than one hundred, outdoor activities should be limited.

The EPA provides information about air quality online. You can also find the Air Quality Index in the local newspaper. Sometimes, people can find out the Air Quality Index on the local news when meteorologists talk about the weather.

Air Quality Index

AQI Numbers	Colors
0 to 50	Green
51 to 100	Yellow
101 to 150	Orange
151 to 200	Red
201 to 300	Purple
301 to 500	Maroon (usually not shown)

Green indicates good quality air. Dark maroon indicates hazardous air quality. The orange range is unhealthy for sensitive groups, such as people with lung disease or other breathing problems.

People Can Reduce Air Pollution

The best way to reduce air pollution is to prevent more pollutants from getting into the air. Limiting car exhaust fumes, fires, and factory emissions can make a big difference in air quality.

Scrubbers: One way to keep chemicals out of the air is by putting filters over or in smokestacks. These filters are called scrubbers. They "scrub" the air, removing chemicals and materials such as ash. The air that is released from the smokestack of the factory contains water vapor and a small amount of pollutants.

Scrubbers can remove up to ninety percent of pollutants before they get released into the air. One drawback of scrubbers is their high cost.

Clean Energy: Another way to improve air quality is to use clean energy. You may have heard of the term *clean energy*. **Clean energy** is a source of energy that does not release pollutants when it is used.

> **Vocabulary**
>
> **clean energy, n.** an energy resource that does not release pollutants when it is used

A clean energy source that people use more and more is solar energy. Solar energy comes from the sun. People in many places around the world now use solar panels to capture light from the sun and turn it into energy.

Another clean energy source is wind. In windy places, people build wind turbines to capture energy from the wind and turn it into electrical power.

Using more clean energy sources helps reduce the amount of pollutants that enter our atmosphere.

By replacing nonrenewable energy with clean energy, people prevent pollutants from entering the air.

Land Contamination

Chapter 5

Big Question

What is land contamination?

Can you think of some ways that this land may have become contaminated?

If you were walking along a trail and saw a sign like this, what would you think? What is **contaminated land**? It is soil that contains chemicals that are harmful to the environment and to people. There are many ways that land can become contaminated.

Vocabulary

contaminated land, n. land that contains chemicals that are harmful to the environment

Land contamination can happen naturally as when ocean salt water floods farmland. It can be accidental if something such as a chemical or oil spill occurs. Or it can be intentional if waste from factories or homes is disposed of improperly.

Garbage Has to Go Somewhere

Once or twice a week, most households put their garbage cans or bags near the street. But where does it all go?

Communities have two ways for garbage to be removed. One is through a recycling company. However, only about thirty percent of garbage is recycled. The rest is waste that cannot be recycled. Some of the waste may be **biodegradable**. This waste is dumped into a **landfill**. A landfill is a place where waste is buried underground. No matter where you live, there is probably a landfill somewhere in your area.

There are three ways to cut down on the amount of waste thrown away: reduce, reuse, or recycle. Reducing how much you use, such as by using a metal water bottle rather than single-use water bottles, reduces landfill waste. Reusing some materials, such as donating old clothes, reduces landfill waste. Recycling materials, such as metal cans and paper, reduces landfill waste.

Vocabulary

biodegradable, adj. describing material that decays through the action of bacteria or other living organisms

landfill, n. a place where waste is buried underground

Recycling	Landfill

A Good Landfill Uses Technology

A landfill must be properly built to prevent contaminated land.

First of all, where the landfill is placed is really important. Landfills should never be located near water that can flood.

Landfills must also be carefully constructed so they don't leak waste materials. To build a solid landfill, construction workers dig a large hole and place thick layers of clay on the bottom. Thick plastic covers the clay. The clay and plastic liner will prevent materials from the garbage from leaking into the soil around the landfill.

At the bottom of the plastic liner, pipes collect liquids released by garbage as it breaks down. The collected liquid moves through the pipes to a wastewater treatment plant, where it is cleaned.

On top of the pipes is a layer of sand and gravel. The liquids filter through these materials before entering the pipes.

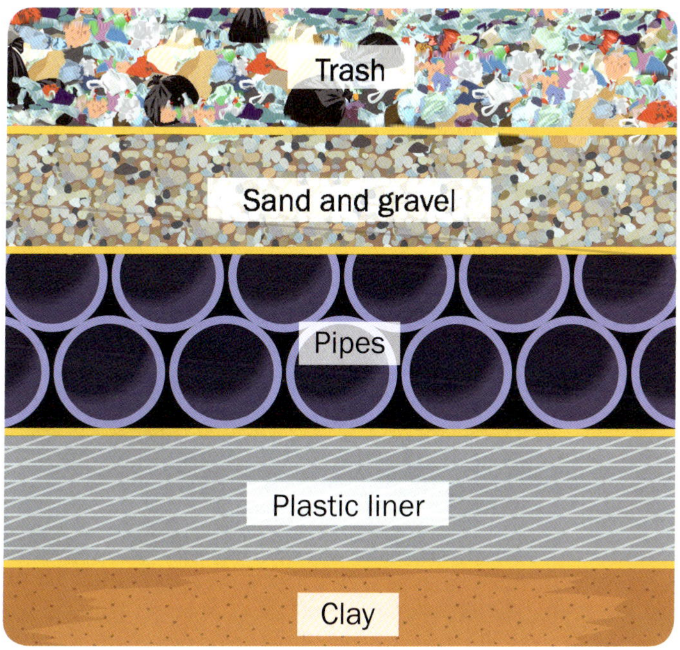

Good Landfills Can Become Green Space Again

When the landfill is securely built, garbage trucks dump the trash they collect into it. The trash is spread out in a layer. A layer of soil is spread on top of the layer of trash. The layers continue to pile up until the landfill is full. It can take about fifty years for a landfill to become full.

After the landfill is filled, it is covered with layers of soil and left alone. After several years, the land can be used for trails and parks.

Putting trash into properly built landfills prevents land from becoming contaminated with the waste products that are breaking down from the garbage that settles there.

People walking on this trail may never guess that layers of trash are safely buried beneath the soil!

Living Off the Land

Chapter 6

Some animals make their homes above ground, in trees, in grasslands, in rocks, and even in the walls of buildings. Some rabbits, earthworms, chipmunks, ants, moles, and many other animals dig burrows in the ground for shelter. Burrows keep animals safe from predators. One way or another, animals depend on land for shelter.

Big Question

How can human activity contaminate land?

Interactions with nonliving elements of land, including rocks, soil, minerals, and nutrients, are essential for growing food and providing shelter for living things to survive.

What will happen if the turtle burrow were filled in and buildings were put in its place?

People Use Land for Important Activities

People must be careful about how they use the land. If they do not use care when building, farming, mining, or using land in other ways, they can cause damage. Damaged land is not easy to fix. Damaged land may be dangerous for people or other living things.

Building: Before people build on land, a **survey** should be done. A survey is when land is observed, measured, and mapped. In a land survey, the living and nonliving things are identified and evaluated. Special technology helps people survey the surface of the land. It is a good idea to find out what is under the ground, too. Devices such as radar and lasers can help people find out what is located underground.

Vocabulary

survey, n. a determination of land boundaries

This stand has tools needed to do a land survey. The surveyor can use the tools to measure the size and height of the land. She can also find out what is underground using the tools on the stand. She may take photos of the land and any living things on it.

It is never a good idea to build on land without doing a land survey. Protected animals or sensitive plants may live on the land. These animals must be relocated, or a new building site must be chosen. The ground below may contain a cave. If so, building a new structure above a cave would not be sturdy and may collapse over time.

Farming: Farmers depend on land to grow crops or raise animals. People and animals depend on these crops for food. Over many years, farmers have learned ways to take care of the land they depend on.

As plants grow in soil, they absorb nutrients needed for growth. If farmers grow the same type of crop in the same field, over time, the field will no longer contain the nutrients that type of crop needs. However, **crop rotation** fixes this problem! When plants that need different types of nutrients are grown in a field in turns, the nutrients do not get used up, and soil stays healthy.

Pulling up a crop means the roots are no longer there to hold the soil in place. **No-till farming** fixes this problem. Farmers harvest the needed part of the plant and leave the rest in the soil. This helps keep the soil in place. Also, nutrients are added to the soil as the roots break down.

Vocabulary

crop rotation,
n. the practice of changing the type of crop grown in a field each season

no-till farming,
n. the practice of harvesting only the part of the plant needed for food and leaving the rest in the soil

You can see where the previously harvested crops were left to hold down the soil. The new crops are planted around them.

Mining: People mine land to extract natural resources. Fuel such as coal and oil, rocks for building, and precious stones and minerals such as gold, copper, and salt are mined around the world.

All types of mining change Earth's surface. Some mines are not very deep. Other mines go hundreds of feet below Earth's surface. Some types of mining contaminate land more than others. In strip mining, topsoil is removed so miners can reach and remove rocks below it. When material is removed from inside or on top of Earth, it leaves massive holes, and the land's usefulness for growing food or providing shelter is limited.

Mining does not just remove materials. It also leaves behind waste. For example, miners take minerals out of certain types of rock, dig up the rocks, remove the minerals they want, and leave the rest. The materials left behind may contain chemicals that are dangerous. This waste material adds up and causes land contamination.

You can see how layers of land have been stripped away so this mine can remove materials from underground.

More People Means More Land Use

People use land for many other reasons. As you drive through your neighborhood, you may see land being cleared to make room for houses, apartments, malls, or office buildings. As the population of people grows, the need for more places to live and work grows with it.

People need to be able to get from one place to another. So, land is cleared to build roads and highways. Airports and railroads require land to be cleared also. When land is used by people, the animals that live there must move to new places. Some animals can adapt to a new environment, but some do not survive their relocation.

This land is being cleared to build a road. What will happen to the things living here?

People Can Preserve Natural Spaces

The government zones, or designates, land for different purposes. Some land is zoned for building. Some land is zoned for farming. Other land is zoned for green space. **Green space** is land that must be left alone. It is set aside for people to enjoy.

Lands that are set aside by the federal government are made into national parks. Lands that are set aside by the state government are made into state parks. Local governments and neighborhoods also set aside land for parks, trails, and wildlife preserves.

Vocabulary

green space, n. land that is set aside and left without buildings

Have you ever visited a national park or a state park? You may have one near where you live!

Sharing the Environment

Chapter 7

Earth is a system made up of interacting parts.

The geosphere is the solid parts of Earth's structure. The hydrosphere includes all of Earth's water. The atmosphere is the layer of air that surrounds the planet. And the biosphere is all the living organisms on Earth.

Big Question

How do human activities that affect water, air, and land impact ecosystems?

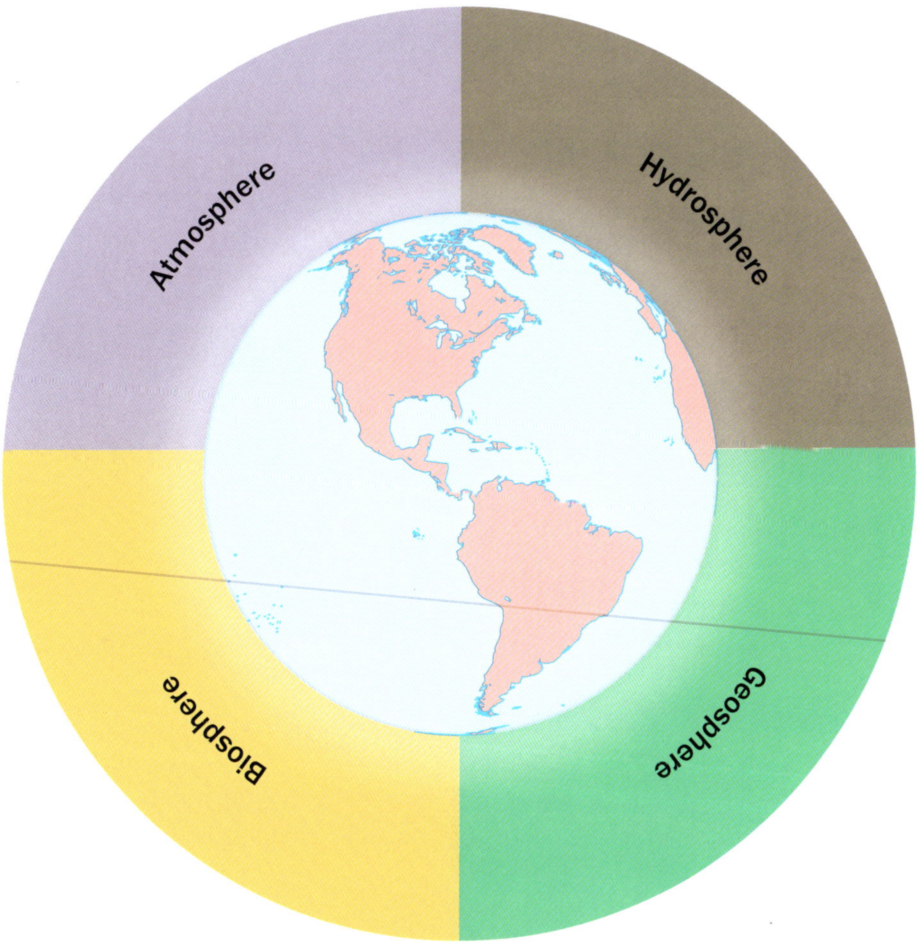

All of Earth's systems interact with each other.

Earth's Spheres Interact

Earth's spheres interact, both directly and indirectly. Events that take place in one of Earth's spheres can affect the other spheres. Look at this photo of a volcano erupting. When a volcano erupts, rocks melt and change. The land changes. A volcano is part of the geosphere.

However, when a volcano erupts, other spheres change also. Lava will burn a path through plants and trees as it flows. Animals and people that live nearby must flee. These are the living parts of Earth, the biosphere. Can you identify any other spheres that could be affected by a volcanic eruption?

Volcanic eruptions affect all of Earth's spheres.

Spheres Interact in Ecosystems

Ecosystems are different types of environments that form from sphere **interactions**. Ecosystems are always changing due to the interactions of the parts within them. These changes can be positive or negative. Plants, animals, and people can cause ecosystems to change.

Vocabulary

ecosystem, n. all the living and nonliving things that interact in a given area

interaction, n. a relationship in a system through which factors affect each other

biodiversity, n. the variety of species on Earth or in any one environment on Earth

Plants Can Disrupt Ecosystems

The more different types of plants in an ecosystem, the more types of animals that live there. The amount of different plants and animals living in a place is known as **biodiversity**.

Biodiversity can be disrupted. The Brazilian pepper tree grows in any type of soil and can take over an area. There are no native animals that eat the plant. As it grows, it takes over land where native plants grow. This throws off the balance of the ecosystem.

Brazilian pepper trees are native to South America. In North America, they can outgrow native North American plants and disrupt the ecosystem they are in.

Animals Can Disrupt Ecosystems

Coral reefs are sensitive ecosystems. If a lionfish moves to a coral reef, the reef becomes endangered. This is because lionfish throw an ecosystem out of balance. Although lionfish look beautiful, they cause major damage.

Lionfish have no predators on coral reefs. They reproduce quickly, laying up to two million eggs a year. Lionfish eat any animal in sight, causing a lack of food for other living things on the reef. They especially like to eat grazing fish. Grazing fish eat algae that grow on coral reefs. Without grazing fish, there could be an overgrowth of algae, which could cause the coral to die. The only way to rid coral reefs of lionfish is for divers to remove them one by one.

Lionfish have dramatic coloring and fins. Lionfish are venomous, and their stings are extremely painful. They can cause damage to coral reefs.

People Can Disrupt Ecosystems

In the past, groups of people used to move from place to place as a way of trying to avoid using up the resources of the land. Today, most people live in one place their entire lives. Some have little regard for how their behaviors affect the planet. As such, people can affect ecosystems both positively and negatively.

New trees start to grow where a forest fire has been, but it will take years for this ecosystem to recover from the disruption.

Wildfires: One way that people change ecosystems is through fire. If a campfire is left burning or sparks from it spread, an entire area could burn down. In a forest fire, houses, trees, and debris are burned. Animals and people must flee. Smoke can pollute the air and cause breathing problems. All of Earth's spheres are affected by a fire. Ash and debris can contaminate nearby water sources. The composition of the soil changes.

Controlled Burns: Sometimes, forest rangers burn parts of a forest on purpose. This is a disruption of the ecosystem, so why would they do this? Scientists have found that not all effects of disruption are negative. Fires do occur naturally in some ecosystems. For instance, fire is especially helpful to certain pine forests. Pine needles take a very long time to decompose. After the pine needles, leaves, and other debris have been burned in a pine forest, there is room for new trees to grow. The fire returns nutrients to the soil, so new plants grow more quickly. By setting planned and controlled fires, park rangers help some ecosystems become stronger.

Some types of trees have cones that are completely sealed with resin, such as the pine cones in the photo on the left. When fire occurs, the resin melts, and the cone opens and releases seeds. In the photo on the right, you can see the places where seeds fell out. The seeds will grow quickly in the nutrient-rich soil.

People Can Protect Ecosystems from Negative Changes

Since Earth's spheres interact, a change in one sphere can change the entire ecosystem. Changes to nonliving parts of ecosystems—the land, water, and air—change the living parts. People cannot stop changes in ecosystems, but they can try to prevent some of the negative ones that take place.

Monitoring: Monitoring is a way to collect data to measure how healthy an ecosystem is. Scientists monitor parts of ecosystems in many ways. For instance, they can count the number of living things or use GPS tracking technology to learn more about specific parts of an ecosystem. Scientists measure soil, water, and air quality using a variety of instruments. They can also input the data they collect into a computer that helps them analyze information, look for patterns, and make predictions about how the ecosystem is changing.

This bee is carrying a tiny GPS tracking device. The device does not harm the bee. Scientists track the bee using a drone. They are trying to find out why populations of bees are declining.

Protection: When scientists analyze data and predict that an ecosystem is experiencing a negative change, they help create laws to protect it. Ecosystems can be protected by keeping people away from certain parts of the environment. For instance, it may be illegal for people to visit certain parts of the land or build on it. Ecosystems can also be protected by passing laws to make it illegal for people to be near certain animals.

In the 1960s, scientists who monitored bald eagles found that the population was declining. They found that the eagle eggs were fragile and fewer chicks were being born each year. The scientists investigated and found that a chemical used to kill pests was impacting the ecosystem. Laws were passed so that the chemical could no longer be used. Due to these laws, populations of eagles have recovered.

DDT, a chemical that made eagle eggs weak, nearly caused eagles to become extinct. Laws preventing the use of DDT helped the eagle population to recover. Other laws also protect eagles. It is illegal to harm an eagle or to even have an eagle feather.

Glossary

A

air pollution, n. any type of substance that is in air and makes it unsafe to breathe (15)

air quality, n. a measure of how clean the air is (22)

algal bloom, n. rapid growth of microscopic algae in a body of water (6)

B

bacteria, n. tiny organisms that break down matter (3)

biodegradable, adj. describing material that decays through the action of bacteria or other living organisms (28)

biodiversity, n. the variety of species on Earth or in any one environment on Earth (39)

C

clean energy, n. an energy resource that does not release pollutants when it is used (26)

conserve, v. to save or protect (9)

contaminated land, n. land that contains chemicals that are harmful to the environment (27)

crop rotation, n. the practice of changing the type of crop grown in a field each season (33)

D

disinfection, n. the process of cleaning infection from a material (3)

E

ecosystem, n. all the living and nonliving things that interact in a given area (39)

F

fresh water, n. water that is usable by humans for things such as drinking and cleaning (8)

G

green space, n. land that is set aside and left without buildings (36)

groundwater, n. water stored in the spaces between materials beneath Earth's surface (8)

I

interaction, n. a relationship in a system through which factors affect each other (39)

L

landfill, n. a place where waste is buried underground (28)

N

natural resource, n. materials that occur in nature that can be used by people (11)

no-till farming, n. the practice of harvesting only the part of the plant needed for food and leaving the rest in the soil (33)

O

ozone, n. a gas found in air (23)

P

pollutant, n. an artificial or natural substance that contaminates air, water, or soil (10)

problem, n. a want or a need (4)

S

smog, n. a type of air pollution that results from a mixture of gases from burning fuels, air, and sunlight that makes air look hazy (16)

surface water, n. fresh water found in rivers, streams, and lakes on Earth's surface (8)

survey, n. a determination of land boundaries (32)

W

wastewater, n. water that has been used by people (2)

water pollution, n. the presence of harmful substances or matter in water (10)

water quality testing, n. methods to measure the types and amounts of chemicals in water (5)

water treatment plant, n. a facility that cleans wastewater before it is discharged into the environment (2)

Core Knowledge®

CKSci™

Core Knowledge Science™

Series Editor-in-Chief

E.D. Hirsch Jr.

Editorial Directors

Daniel H. Franck and Richard B. Talbot

Subject Matter Expert

Lisa M. Landino, PhD
Garrett-Robb-Guy Professor,
Chemistry Department
College of William and Mary
Williamsburg, VA

Illustrations and Photo Credits

Aleksandr Proshkin / Alamy Stock Photo: 4
ALEXEY KORNYLYEV / Alamy Stock Photo: 35
All Canada Photos / Alamy Stock Photo: 33, 41
Beara Creative / Alamy Stock Photo: i, iii, 40
Dan Leeth / Alamy Stock Photo: 36
David Bokuchava / Alamy Stock Photo: 20b
dbimages / Alamy Stock Photo: Cover B, 22a
Dmitry SHISHKIN / Alamy Stock Photo: 12a
dpa picture alliance / Alamy Stock Photo: 12b
Eugenie Robitaille / Alamy Stock Photo: 42a
Frank Bienewald / Alamy Stock Photo: 26b
gary corbett / Alamy Stock Photo: 39
incamerastock / Alamy Stock Photo: 17
James King-Holmes / Alamy Stock Photo: Cover A, 43
Jan Donath / Alamy Stock Photo: 34
Jane Haslam / Alamy Stock Photo: Cover C, 22b
Jeff Zehnder / Alamy Stock Photo: 25
Jim McKinley / Alamy Stock Photo: 11
makasana photo / Alamy Stock Photo: 14a
Manit Larpluechai / Alamy Stock Photo: 32
Martin Smart / Alamy Stock Photo: Cover E, 44
Marvin Dembinsky Photo Associates / Alamy Stock Photo: 42b
Matjaz Corel / Alamy Stock Photo: 20a
Michael Klinec / Alamy Stock Photo: 13a
Mira / Alamy Stock Photo: 30
Olli Geibel / Alamy Stock Photo: 15
Panther Media GmbH / Alamy Stock Photo: 6
Paul Heaton / Alamy Stock Photo: 26a
Rapeepong Puttakumwong / Alamy Stock Photo: 10
Rawf8 / Alamy Stock Photo: 1
RBM Vintage Images / Alamy Stock Photo: 23
Riccardo Sala / Alamy Stock Photo: 18
riddypix / Alamy Stock Photo: Cover D, 27
Stocktrek Images, Inc. / Alamy Stock Photo: 14b
trekandshoot / Alamy Stock Photo: 16
Tristan3D / Alamy Stock Photo: 7
Valery Voennyy / Alamy Stock Photo: 5
Westend61 GmbH / Alamy Stock Photo: 38
wonderlandstock / Alamy Stock Photo: 13b
Zoe17es / Alamy Stock Photo: 8